Measurement Match

Match each person with the sentence that best

__h__	The 35-millimeter film was developed in 2 hours.	a. Olympic swimmer
_____	Tara ran her best in the 1500-meter race.	b. art historian
_____	Pedro needed a 15-millimeter wrench to repair the car motor.	c. gardener
_____	The 50-meter pool opens for practice at 5 A.M.	d. jewelry designer
_____	The shortest trees in the garden still grew to be 2 meters high last year.	e. geologist
_____	The incision to remove Tim's appendix was 7 centimeters long.	f. politician
_____	Janet collected rock samples at the bottom of a canyon that was 0.23 kilometer deep.	g. runner
_____	Terry walked 6 kilometers passing out campaign flyers for his state senate campaign.	h. photographer
_____	The width of the fake western art print was 33 centimeters.	i. car mechanic
_____	The red glass bead was 4 millimeters in diameter.	j. surgeon

Measuring Length

A **measurement** is a *number* followed by a *unit*. Here are four examples of possible measurements of length.

2 pencil lengths   2 feet   6 book widths   4 car lengths

A **unit of length** can be anything that has a fixed length. Which of the following could be used as a unit of length? Write "yes" or "no" next to each one.

_____ dog barks   _____ hand spans   _____ paper clips

_____ star light   _____ rubber bands   _____ shoes

_____ desks   _____ colors   _____ jokes

Now finish this table.

	Number	Unit of Length
The ramp was 20 skateboards long.	20	skateboards
The movie theater was 25 paces wide.		
The ribbon was 4 hand spans long.		
Tina rode her unicycle 15 blocks.		

You have some units, like those shown below, with you all the time.

thumb width   hand span   cubit   pace

Use your thumb width to measure these objects.

The height of this page is about _____ thumb widths.

The length of the picture of the bus is about _____.

Did you remember to write the unit of measure with each answer? ☐ Yes ☐ No

Use your hand span to measure these objects. Don't forget to write the unit of measure with your answer.

This book lying open is about _____ hand spans across.

The longest side of your desk or table is about _____.

The height of the door is about _____.

Use your cubit to measure these objects.

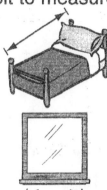

The length of your bed is about _____ cubits.

The width of a windowpane in your room is about _____.

Measure these distances in paces.

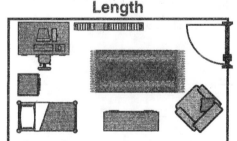

It is about _____ paces from my desk to the door.

My room is about _____ wide.

The distance from the front door to the kitchen is _____ paces.

Team Project #1: Measuring Your World

Your Room

Look around your room. There may be floor tiles, ceiling tiles, boards, or something else that you can use as units to estimate the size of the room.

The unit I choose to estimate the size of my room is an _____.

My room is about _____ _____ long.

My room is about _____ _____ wide.

My room is about _____ _____ high.

Desk and Chair

Decide on a good unit to use to measure your desk and chair. The unit could be a calculator, a card, a pencil, or some other small object.

The unit I choose to measure my desk and chair is an _____.

My desk is about _____ _____ long.

My desk is about _____ _____ wide.

My desk is about _____ _____ high.

The seat of my chair is about _____ _____ high.

The back of my chair is about _____ _____ high.

The back of my chair is about _____ _____ higher than my seat.

Now invent your own original unit of length and measure two distances with it. Write a sentence for each distance that tells what you measured and what unit you used.

_____

_____

A "Good" Unit

How do you decide what makes a "good" unit for measuring a length?

_____

Common Metric Units of Length

In the metric measurement system, the common units for measuring length are **millimeter, centimeter, meter,** and **kilometer.**

Think about these objects to help you visualize the metric units of length.

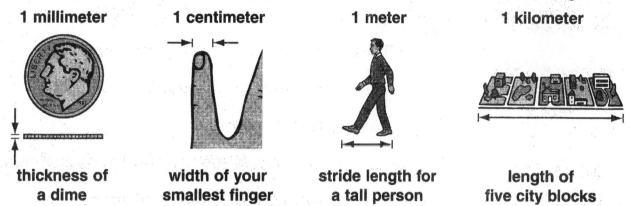

1 millimeter	1 centimeter	1 meter	1 kilometer
thickness of a dime	width of your smallest finger	stride length for a tall person	length of five city blocks

Match each measurement with the appropriate picture.

a. 1 millimeter

b. 1 centimeter

c. 1 meter

d. 1 kilometer

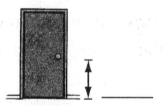
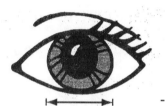
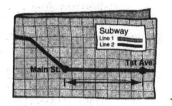
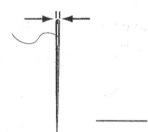

Choose a unit of measure that makes sense for measuring each distance: millimeter, centimeter, meter, or kilometer. (There may be more than one correct answer.)

length of a river _____   length of your thumb _____

depth of the ocean _____   height of a school building _____

thickness of a penny _____   distance you can throw a ball _____

length of a dollar bill _____   length of a butterfly's wing _____

your height _____   distance you cycle in an hour _____

height of a coffee cup _____   length of a swimming pool _____

Making Sense of Cents

Use a dictionary if you need it. Fill in the blanks using the list of words.

The _____ thermometer measures temperature.

Does a _____ really have 100 legs?

One hundred _____ makes one dollar.

A _____ shows parts equal to hundredths.

A _____ is 100 years.

A _____ was the leader of 100 Roman soldiers.

A _____ marks the end of 100 years.

One hundred _____ makes one meter.

What number do all these words have in common? _____

cents
century
centigrade
percent
centimeters
centennial
centipede
centurion

millifold
millipede
millennium
million
milliliters
millimeters

There are 1000 _____ in one meter.

One _____ is one thousand thousandths.

A _____ doesn't really have a thousand legs.

A _____ is one thousand years.

The word _____ means one thousand times.

There are 1000 _____ in one liter.

What number do all these words have in common? _____

Centimeters, Decimeters, and Meters

This square measures 1 **centimeter** on each side. ☐

Which of the following lengths are approximately 1 centimeter long at actual size? Write "yes" or "no" next to each one.

_____ 🖩   _____ 👍   _____ 📎

Name four objects in the room that are approximately 1 centimeter long.

_____   _____   _____   _____

The strip at the bottom of the page is 10 centimeters long. This length is equivalent to 1 **decimeter**.

Which of the following lengths are approximately 1 decimeter long at actual size?

_____ 🖩   _____ 👍   _____ ON/CE

Name four objects in the room that are approximately 1 decimeter long.

_____   _____   _____   _____

Using ten 1-decimeter strips, you can make a strip that is 10 decimeters long. This length is equivalent to 1 **meter**.

Which of the following lengths are approximately 1 meter long?

_____ 🐕   _____ ⌨️   _____ 🛹

Name two objects in the room that are approximately 1 meter long.

_____   _____

©2000 by Key Curriculum Press
Do not duplicate without permission.

Measuring in Centimeters

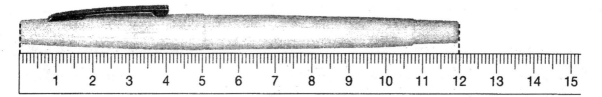

How many centimeters long is the picture of the pen? _____ centimeters

To measure an object, line up the **zero mark** on the ruler so that it is even with one end of the object.

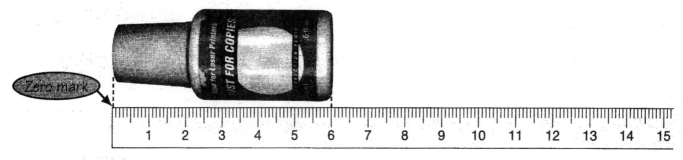

The bottle is 6 centimeters long.

Sometimes the zero mark is at the end of the ruler, sometimes not.

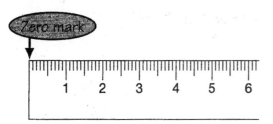   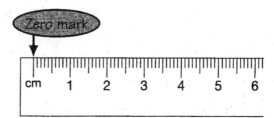

Which rulers are lined up correctly to the paper clip?

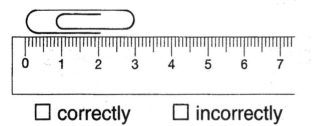

☐ correctly   ☐ incorrectly

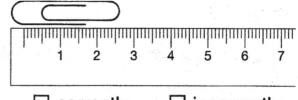

☐ correctly   ☐ incorrectly

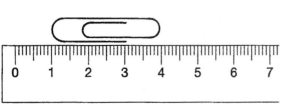

☐ correctly   ☐ incorrectly

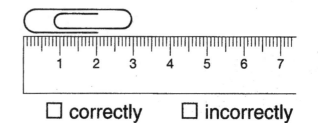

☐ correctly   ☐ incorrectly

How long is the picture of each object below?

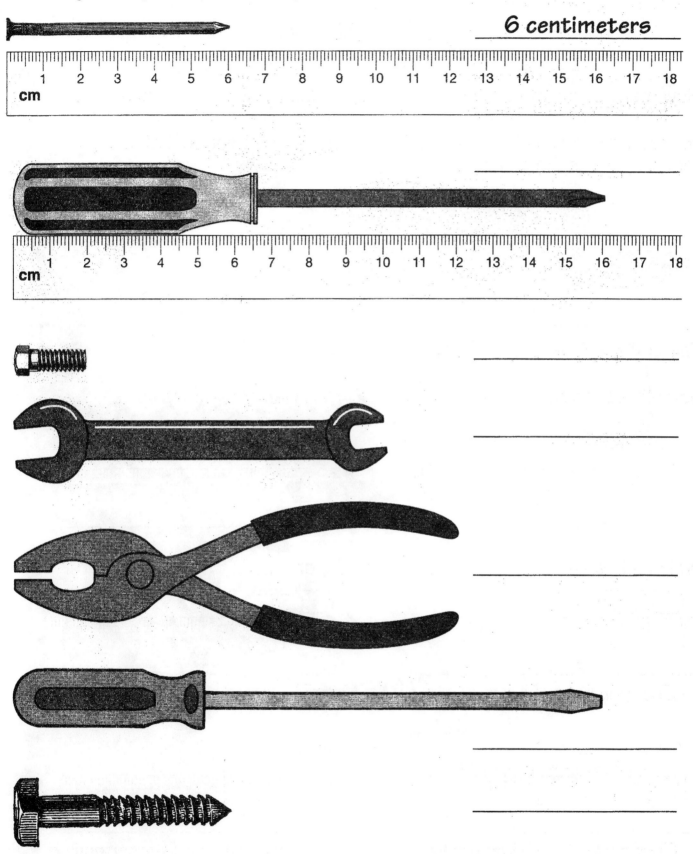

6 centimeters

Did you remember to write "centimeters" with each answer? ☐ Yes ☐ No

Measure the length of each picture in centimeters using a ruler or a 20-centimeter tape (#1 or #3) on page 45. Write each length in the circle shown. You can write "cm" instead of "centimeters" with each measure.

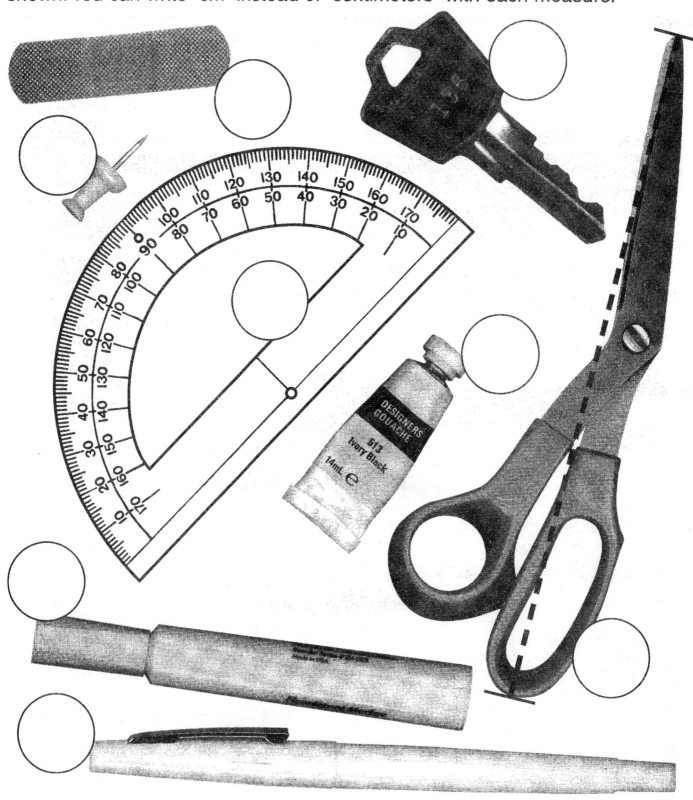

Did you remember to write "cm" or "centimeters" in each circle? ☐ Yes ☐ No

Team Project #2: Using Your Hand Span as a Measuring Device

Suppose you are out in the woods or in the street and you need to measure something. If you don't have a ruler, one thing you can use is your hand.

Step 1 Measure your hand span to the nearest centimeter.

My hand span is _____ centimeters.

Step 2 Measure the height of another member of your team to see how tall he or she is.

My teammate is _____ hand spans tall.

Step 3 Multiply the two numbers to get the number of centimeters.

_____ × _____ = _____

   number of   the measure of   measurement
   hand spans   your hand span   of height
                         in centimeters   in centimeters

My teammate's height is _____ centimeters.

Now use your hand span to measure the length of four objects of your choice.

Name of object	Number of hand spans	Measure of your hand span in centimeters		Actual length in centimeters
_____	_____	× _____	cm =	_____ cm
_____	_____	× _____	cm =	_____ cm
_____	_____	× _____	cm =	_____ cm
_____	_____	× _____	cm =	_____ cm

©2000 by Key Curriculum Press
Do not duplicate without permission.

Ruler Rejects

Measure the length of this picture of a toothbrush.
It is _____ centimeters long.

Now look carefully at each ruler below. Why would each ruler cause problems if you used it to measure the length of the toothbrush?

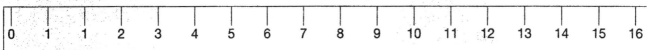

What's wrong with this ruler? _____

_____

If you used this ruler to measure the toothbrush, how long would it say the toothbrush was? _____

What's wrong with this ruler? _____

_____

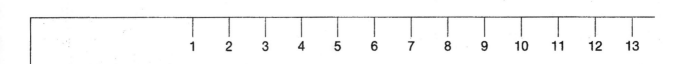

What's wrong with this ruler? _____

_____

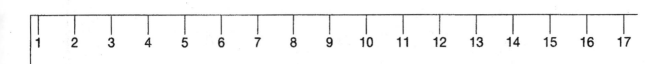

What's wrong with this ruler? _____

_____

Measuring to the Nearest Centimeter

When the length of an object falls between two centimeter marks on a ruler, you can measure its length to the *nearest centimeter*.

Choose the measurement for each picture that is correct to the nearest centimeter.

The length of the caterpillar is nearest to 8 cm.

☐ about 8 cm
☐ about 7 cm

☐ about 8 cm
☐ about 9 cm

☐ about 9 cm
☐ about 10 cm

Measure the length of each caterpillar pictured below to the nearest centimeter.

about 5 cm

"5" is the closest centimeter mark.

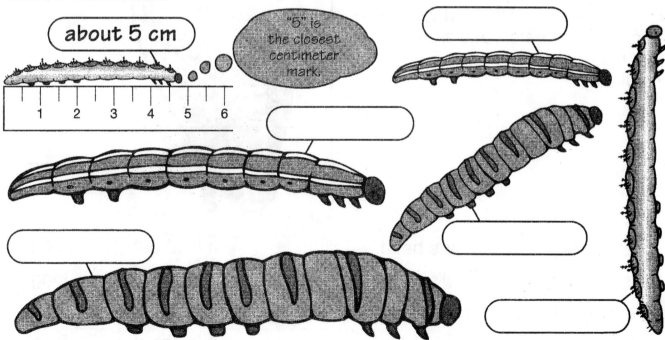

Did you remember to write the unit of measure with each answer?   ☐ Yes   ☐ No

Measuring a Line Segment to the Nearest Centimeter

A piece of a line with two endpoints is called a **line segment.** The segment below is line segment AB because the two endpoints are labeled A and B. You can write line segment AB as $\overline{AB}$. This line segment can also be called line segment BA and can be written as $\overline{BA}$.

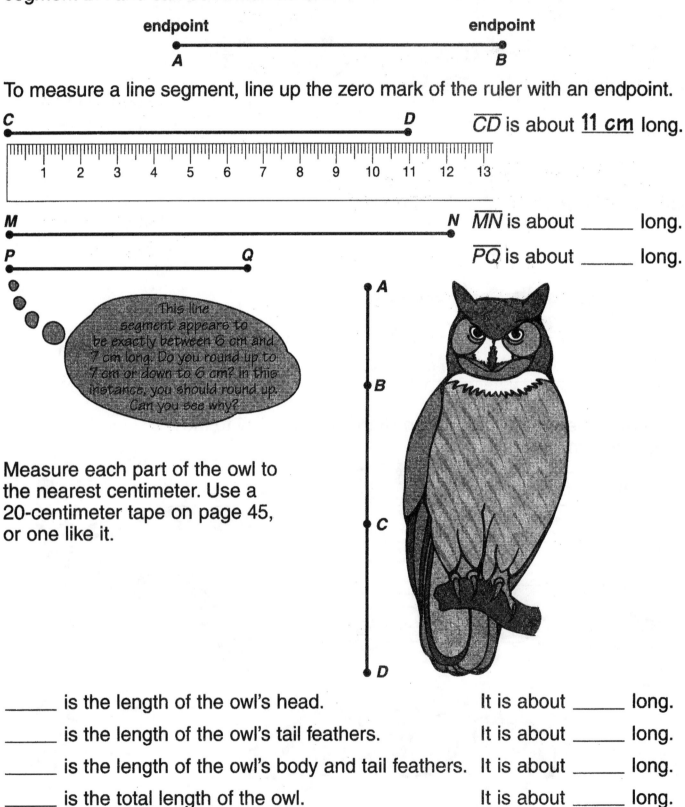

To measure a line segment, line up the zero mark of the ruler with an endpoint.

$\overline{CD}$ is about __11 cm__ long.

$\overline{MN}$ is about _____ long.

$\overline{PQ}$ is about _____ long.

This line segment appears to be exactly between 6 cm and 7 cm long. Do you round up to 7 cm or down to 6 cm? In this instance, you should round up. Can you see why?

Measure each part of the owl to the nearest centimeter. Use a 20-centimeter tape on page 45, or one like it.

_____ is the length of the owl's head.   It is about _____ long.

_____ is the length of the owl's tail feathers.   It is about _____ long.

_____ is the length of the owl's body and tail feathers.   It is about _____ long.

_____ is the total length of the owl.   It is about _____ long.

Team Project #3: Making a Pendulum Clock

You will need

- a small heavy object to serve as the "bob" for the pendulum
- a piece of string at least 50 centimeters long
- a meter tape (like the one on page 45)
- adhesive tape
- a timer to measure seconds. This could be a stopwatch, a clock, or a watch.

Set up your pendulum by attaching the bob to one end of the string. Tape the other end of the string to the edge of a desk or table so that the pendulum can swing freely back and forth.

During this project, you will find the length of string required for the pendulum to complete 10 swings in 10 seconds. To complete a swing, the bob must go both back and forth once.

To start, time how long it takes for your pendulum to complete 10 swings. Measure the length of your string from the edge of the desk or table to the center of the bob. This is your pendulum length. Record your results in the first row of the table below.

If your pendulum does not complete 10 swings in 10 seconds, adjust the length of your string, repeat the experiment, and record another set of results. Continue this process until your pendulum completes 10 swings every 10 seconds. Now you have a simple clock!

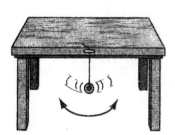

Length of pendulum	Time elapsed for 10 swings

Conclusion 1: One swing takes about _____ second(s).

Conclusion 2: My pendulum is about _____ long.

Conclusion 3: To make my pendulum clock faster, I need to make the string _____.

Meet the Millimeter

Measure the lengths of the two bugs pictured below to the nearest centimeter.

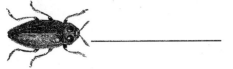   

Both of your measurements were probably the same, but the bugs are different lengths. If you want to measure more precisely, you can use a smaller unit called a **millimeter.** You can write millimeter as mm.

This ruler shows millimeters. The flea is about 1 millimeter long.

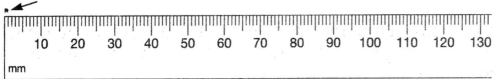

Which of the following objects is about 1 millimeter long? Write "yes" or "no" next to each one.

_____ distance across the head of a pin

_____ length of the eraser at the end of a pencil

_____ length of a bumblebee

_____ thickness of a paper clip

Measure and record the length of each picture below in millimeters. Don't forget the antennae. Use the 20-centimeter tape (#2) on page 45 if necessary.

Locust _____

Housefly _____

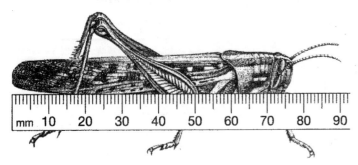   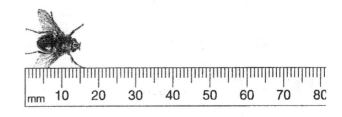

Moving leaf bug

_____

Beetle

_____

Flea

_____

Wood ant

_____

Did you remember to write "millimeter," "millimeters," or "mm" with each answer? ☐ Yes ☐ No

Centimeters and Millimeters

The top ruler shows centimeters and the bottom one shows millimeters.

This ruler shows centimeters.

This ruler shows millimeters.

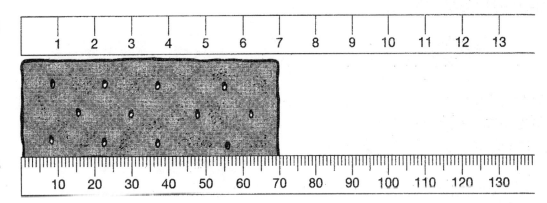

Fill in each blank.

The cracker above is _____ centimeters long.

The cracker above is _____ millimeters long.

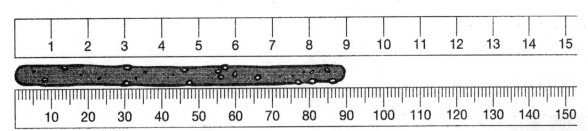

The pretzel stick above is _____ centimeters long.

The pretzel stick above is _____ millimeters long.

This ruler has cm and mm. The numbered lines show centimeters. The shorter, unmarked lines show millimeters.

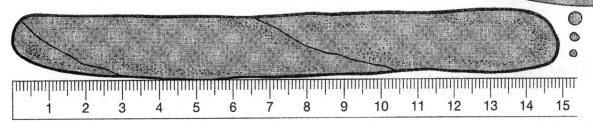

The bread stick above is _____ centimeters long.

The bread stick above is _____ millimeters long.

Fill in each blank.

_____ mm = 1 cm   30 mm = _____ cm

_____ mm = 2 cm   _____ mm = _____ cm

Make It Yourself!

You can make your own measuring tape. Put a ✔ in the ☐ as you finish each instruction.

Turn to page 22 and cut off the bottom of pages 22 and 23 along the line as shown. This will soon be a measuring tape. ☐

Cut off the shaded parts of each strip and throw them away. ☐

The fingerprint on this strip is exactly 1 centimeter wide. Cut it out. ☐

Line up the fingerprint over the end of the paper strip. ☐

To make your 1-centimeter mark, use a pencil to draw a line from the edge of the strip to the arrow on the fingerprint. Label it "1." ☐

Move the fingerprint so it lines up with your new 1-centimeter line and make the 2-centimeter mark. Label it "2." ☐

Continue making centimeter marks to the end of the paper strip. ☐

Compare your measuring tape to a commercial ruler or your 20-centimeter tape and adjust your marks for accuracy if necessary. ☐

Now you will add a mark halfway between each pair of centimeter marks on your measuring tape. The marks on it will be: (Check one.)

☐ 20 millimeters apart ☐ 5 millimeters apart ☐ 10 millimeters apart

Put a ✔ in the ☐ as you finish each instruction.

Make a tiny mark halfway between the zero edge and the 1-centimeter mark on your measuring tape. You can do this by estimating or by folding. Try to put the mark right in the middle. ☐

Now draw the 5-millimeter mark. Don't make it as long as a centimeter mark. ☐

Now draw a 5-millimeter mark between each pair of centimeter marks on your measuring tape. ☐

Measure the length of each picture *to the nearest centimeter.* To measure to the nearest centimeter, you need to have a ruler with 5-millimeter marks so that you can see which centimeter mark you're closest to. Remember that if the length appears to fall exactly on the 5-millimeter mark, you need to make a decision whether to round up or down.

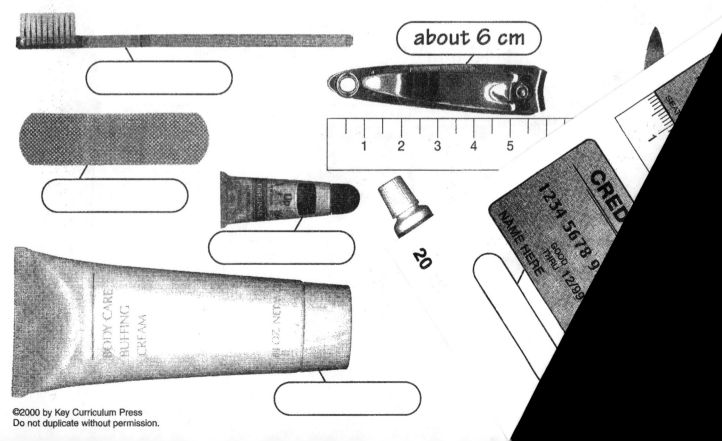

Which does not mean the same as the others?

☐ $\frac{1}{10}$ ☐ 0.1   ☐ 10   ☐ a tenth

Now you will add millimeter marks to your measuring tape. These marks will be: (Check one.)

☐ 0.01 cm apart   ☐ 1 cm apart   ☐ 10 cm apart   ☐ 0.1 cm apart

Put a ✔ in the ☐ as you finish each instruction.

Make four tiny marks between the zero edge and the 5-millimeter mark on your ruler. Try to make them equally spaced. ☐

Now draw the millimeter marks. Don't make them as long as a 5-millimeter mark. ☐

Now draw millimeter marks between each pair of centimeter marks on your measuring tape. ☐

Measure the length of each picture *to the nearest millimeter*. This can be hard, because millimeters are so small!

about 46 mm

Centimeters, Millimeters, and Decimals

There are _____ millimeters in 1 centimeter.

1 millimeter is $\frac{1}{10}$ of a centimeter.

1 millimeter is 0.1 centimeter.

Make a double match.

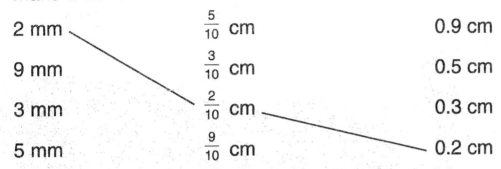

Now let's look at some objects that are longer than 1 centimeter. Fill in the blanks.

The millipede is 2 centimeters plus 3 millimeters long.
The millipede is 2.3 centimeters, or 23 millimeters, long.

The earthworm is 12 centimeters plus _____ millimeters long.

The earthworm is _____ centimeters, or _____ millimeters, long.

Measure and record each length in three different ways.

Object	cm plus mm	cm	mm
length of your shortest finger			
length of your pencil			

The Big Catch

Cut out the 15-centimeter ruler on this page and use it to measure each fish from one end of the dashed line to the other in centimeters. Write your answers in the spaces provided.

15 cm   15 + 15   15 + 15 + 8   15 + 15 + 8 = 38 cm

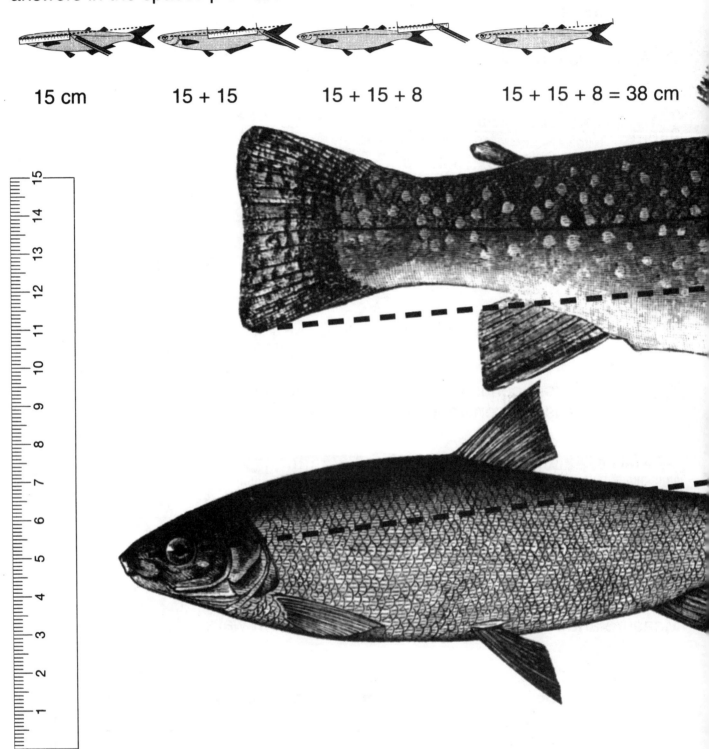

Changing Units

There are different ways to write the measure of a length.

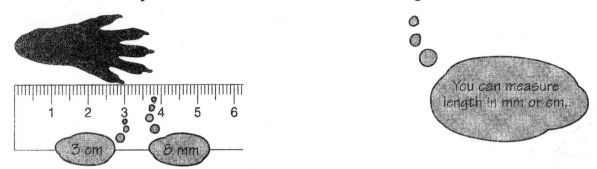

This pack rat footprint is 38 millimeters long.

This pack rat footprint is 3.8 centimeters long.

Measure the length of each picture below in millimeters and then centimeters.

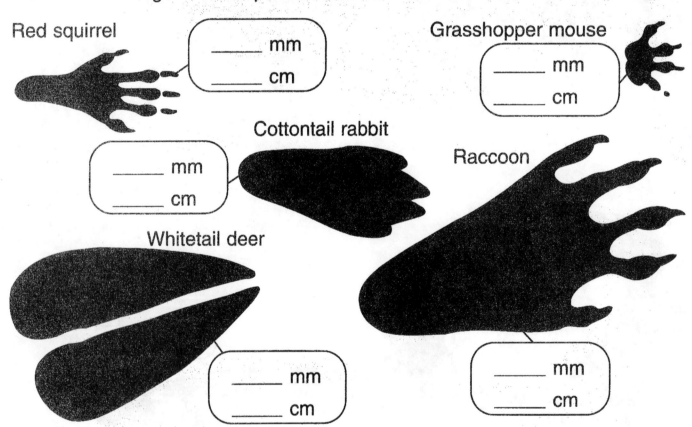

If you know a measurement in millimeters, explain how to write the same measurement in centimeters.

_____

_____

Drawing Line Segments

To draw a line segment that is 6.3 centimeters long, draw a dot centered at the zero mark and one at the 6.3 mark. Then draw a line segment between the dots.

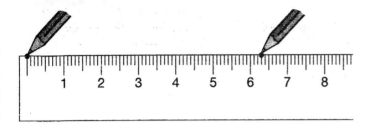 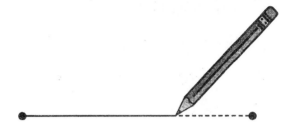

Draw a line segment for each length.

4.7 cm

11.3 cm

2.6 cm

15.8 cm

9.4 cm

0.9 cm

17.2 cm

8.8 cm

Follow the directions to draw each line segment below. The first two examples are done for you.

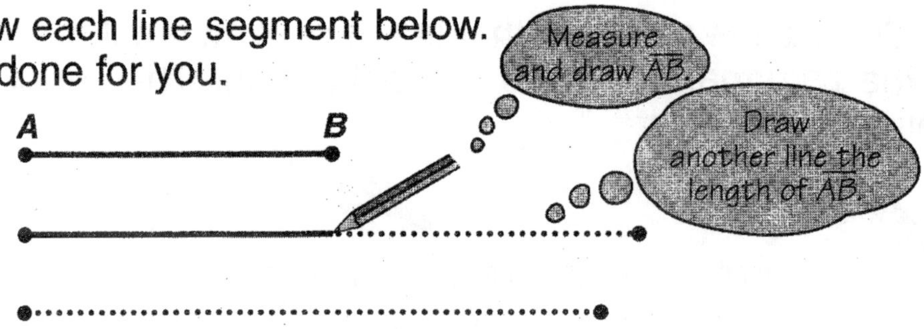

$\overline{AB}$

twice as long as $\overline{AB}$

3.5 cm longer than $\overline{AB}$

$\overline{CD}$

half as long as $\overline{CD}$

4.2 cm shorter than $\overline{CD}$

6.5 cm shorter than $\overline{CD}$

$\overline{EF}$

1.8 cm longer than $\overline{EF}$

2.3 cm shorter than $\overline{EF}$

4.6 cm shorter than $\overline{EF}$

$\overline{GH}$

22 mm longer than $\overline{GH}$

45 mm longer than $\overline{GH}$

12 mm shorter than $\overline{GH}$

37 mm shorter than $\overline{GH}$

half as long as $\overline{GH}$

How Precise Do You Want to Be?

Ruler A, shown below, is a simple ruler with marks that are 1 centimeter apart. It is not a very **precise** measuring tool.

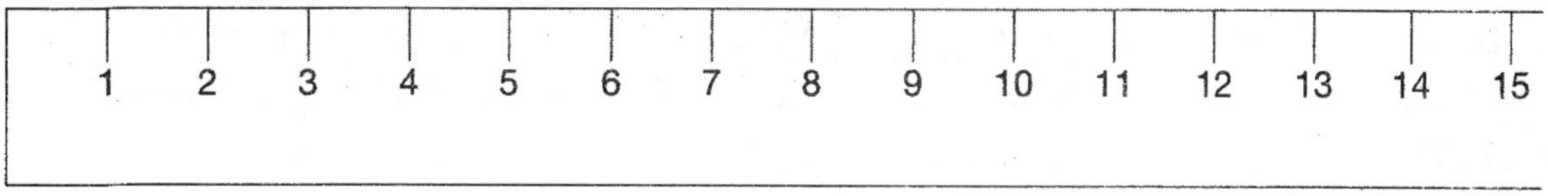

Ruler A

Ruler B is **more precise.** It has marks halfway between each centimeter mark. Draw each missing mark.

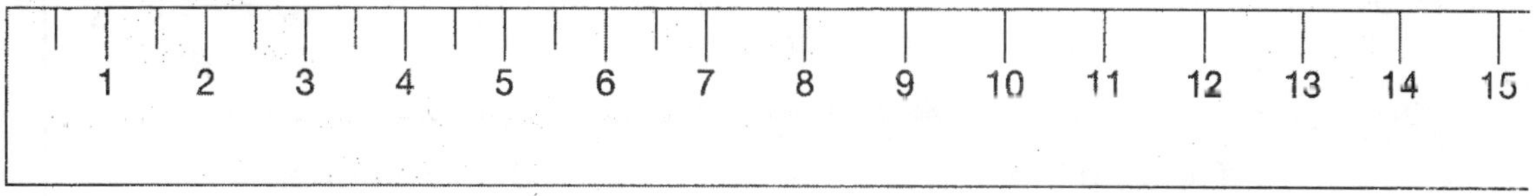

Ruler B

This ruler measures to the
☐ nearest centimeter   ☐ nearest millimeter   ☐ nearest inch

The marks on Ruler C are _____ apart. It is the **most precise** ruler on this page.

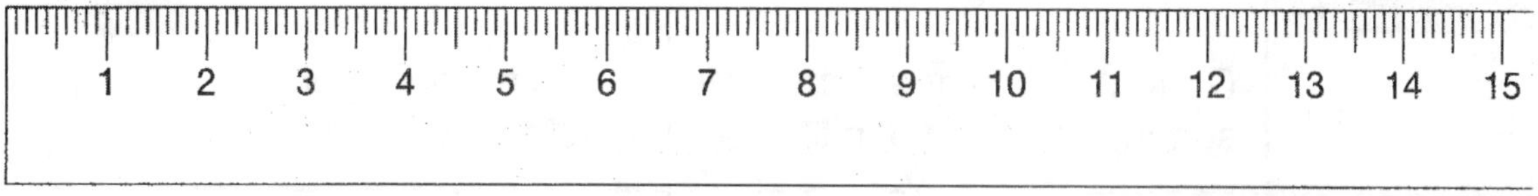

Ruler C

When you measure an object, you must choose your precision. For each situation below, decide how precise you think the measurement needs to be. There could be more than one correct answer!

Sarah measured the width of a patient's tooth for a crown fitting.
☐ to the nearest millimeter   ☐ to the nearest centimeter

Terry measured the length of his cat to tell the veterinarian how large a cat he had.
☐ to the nearest millimeter   ☐ to the nearest centimeter

They measured the floor of her bedroom. She plans to carpet the room and needs to cut out the carpeting correctly.
☐ to the nearest millimeter   ☐ to the nearest centimeter

Sam repairs watches and measured the width of a part from the inside of a customer's watch.
☐ to the nearest millimeter   ☐ to the nearest centimeter

Team Project #4: Scavenger Hunt

To go on a Scavenger Hunt you begin at "start" and solve the clue to find where to go next. Then solve that clue to find the next one. You will visit every clue before you are done. You may need to do some research or to make some measurements in order to solve a clue. Draw a line to show your team's path from clue to clue.

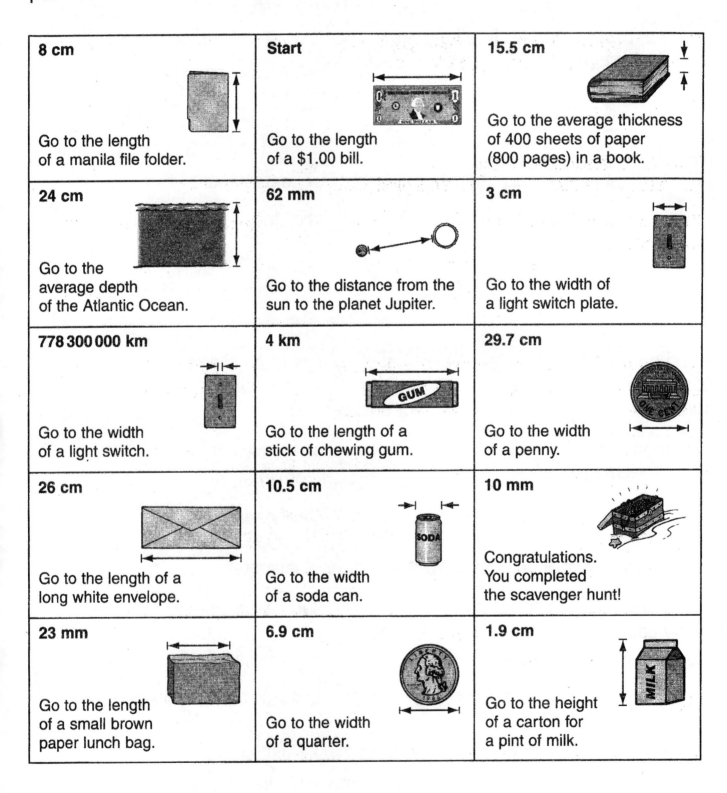

Frog Family

Measure the length of each frog picture to the nearest tenth of a centimeter. Write your answer as a decimal, as shown in the first example below. Make a dot at each end of the length you measure.

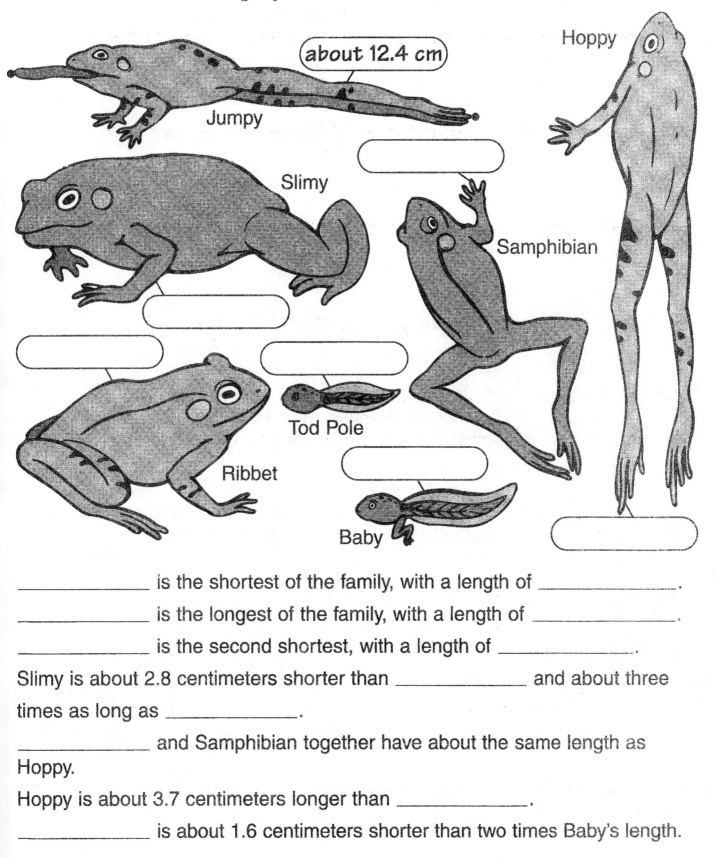

_____ is the shortest of the family, with a length of _____.

_____ is the longest of the family, with a length of _____.

_____ is the second shortest, with a length of _____.

Slimy is about 2.8 centimeters shorter than _____ and about three times as long as _____.

_____ and Samphibian together have about the same length as Hoppy.

Hoppy is about 3.7 centimeters longer than _____.

_____ is about 1.6 centimeters shorter than two times Baby's length.

A Good Eye for Length

Can you draw a 1-centimeter line segment without a ruler? Try it! Draw a 1-centimeter segment in the oval.

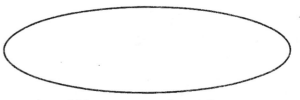

Now get your ruler and measure your line segment. Were you close? Use this space to practice drawing 1-centimeter segments without a ruler.

Now practice drawing a segment 2 centimeters long until you can do it without a ruler.

Now practice drawing a segment 3 centimeters long until you can do it without a ruler.

Now try drawing a line segment that is 6 centimeters long.

Work with a partner and draw segments of lengths 5 centimeters, 10 centimeters, and 20 centimeters without a ruler. Make sure you both agree on the segment lengths. Then check the lengths with a ruler.

Estimate the length of each segment without using a ruler. This is sometimes called **eyeballing**. Write your estimates to the nearest centimeter.

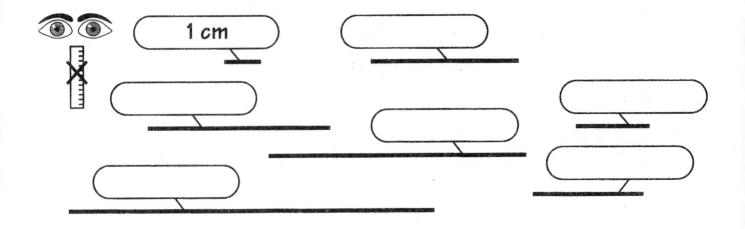

Measure by Eye

Estimate the length of each actual object. Then *measure* each length to see how close you were. Make sure to write cm or mm with each measure.

What to look at	Estimate without a ruler	Measure with a ruler	How far off was your estimate?
length of a paper clip	It's about _____ .	It measures _____ .	My estimate was off by _____ .
length of your pencil			
length of your shoe			
length of your littlest finger			
width of your shoe			
height of your chair			

Measuring Parts of an Object

When measuring only part of an object, line up the edge of your ruler with the part you want to measure.

To measure the face on this wristwatch, line up the edge of the ruler with the edge of the watch face.

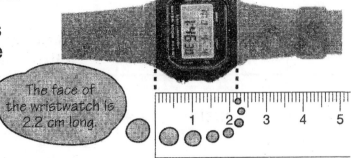

The face of the wristwatch is 2.2 cm long.

Measure each object to the nearest tenth of a centimeter.

outside of frame

_____   _____
 width   height

photo inside the frame

_____   _____
 width   height

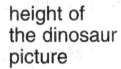

height of the dinosaur picture

_____

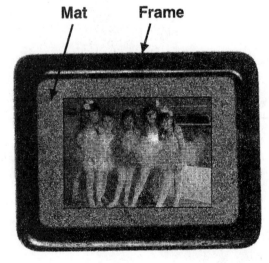

Mat   Frame

height of photo _____

width of photo _____

height of mat _____

width of mat _____

height of outside of frame _____

width of outside of frame _____

32

The images below show the front and back of a Mexican 10 peso bill. Measure each of the following parts of the bill to the nearest tenth of a centimeter.

The length of the bill _____

The width of the bill _____

The height of any letter in the word "Diez" _____

The width of the man's moustache _____

The height of the man standing next to the horse _____

The diameter of the round Banco de Mexico seal _____

Find two other lengths to measure on the 10 peso bill. Describe each and measure it.

_____

_____

Do one of the following:

- Find out who the man with the large moustache is and what he is known for.

- Answer this question: What do you think the two hands holding the corn means?

- Imagine a conversation between the man on the horse and the man standing next to him. Describe what they are saying and why.

Curves

You need a piece of string and a 20-centimeter-long ruler or measuring tape. One way to measure the length of a curve is with a string.

How long would you estimate this curve to be? _____ centimeters

| Now let's measure it. Get some string and cut off a piece that is longer than the curve. ☐ done | Place one end of the string at the end of the curve and then match the string along the curve. ☐ done | Mark the string with a pen or a pencil at the other end of the curve. ☐ done |

Pull the string straight and measure from the end to your mark. ☐ done

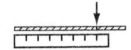   The curve is _____ long.

How did your original estimate compare with your measurement? _____

Why might it be harder to estimate the length of a curve than of a straight line segment? _____

_____

Use string and a ruler or tape measure to measure the length of each snake picture.

_____   _____

On a blank piece of paper, draw a curve that is 30 centimeters long. On the same paper, explain your method for constructing your curve.

34

Measuring Around Objects

To measure the distance around an object, you need a tape measure or a string.

 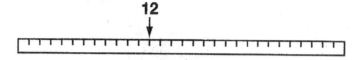

The distance around Gregor's leg at the calf is about 12 centimeters.

Measure each body part as shown below to the nearest centimeter. Use the meter tape you cut from page 45, a tape measure, or a string and a meter stick. You may ask a partner to help you.

around your head _____

around your ankle _____

around your upper arm, with your muscle flexed _____

around your neck _____

around your calf _____

around your wrist _____

Use your measurements to estimate each ratio below.

The distance around my head is about _____ times the distance around my neck.

The distance around my calf is about _____ times the distance around my ankle.

The distance around my wrist is about _____ times the distance around my flexed arm.

Estimating Meters

A meter is about the length of a baseball bat or the length of a guitar.

Find another common object or distance that is roughly a meter long. Draw a picture of this object or distance here.

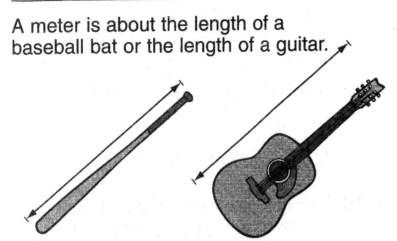

Estimate the length of each object in the first column by deciding if it is shorter than 1 meter, the same as 1 meter, or longer than 1 meter. Check the appropriate box. You don't need to measure every object.

Object	Shorter than 1 meter	About 1 meter	Longer than 1 meter
length of a car			
length of a bed			
height of a bicycle			
length of a skateboard			
your height			
height of a doorknob from the floor			
width of a computer screen			

Find three different lengths on your body that are approximately 1 meter long and write a sentence for each one. (For example: "From my ____ to my ____ is ____." Or "My _____ is ____ long.") Use a measuring tape if necessary.

_____

_____

Estimating Kilometers

The prefix *kilo* means 1000.

The symbol for kilometer is **km**.

Check the most reasonable answer to each question.

How many meters are in a kilometer? ☐ 10 ☐ 100 ☐ 1000

How long does it take to walk 1 kilometer?
☐ 1 minute ☐ 15 minutes ☐ 15 hours

How many city blocks is 1 kilometer? ☐ about 5 ☐ about 25 ☐ about 100

How far is your home from your school?
☐ less than 1 kilometer ☐ about 1 kilometer ☐ greater than 1 kilometer

Circle the best estimate.

length of an adult king cobra snake
5 cm   5 m   5 km

height of Mount Everest
9 cm   9 m   9 km

length of a marathon
42 cm   42 m   42 km

length of a sewing needle
4 mm   4 cm   4 m

length of the St. Lawrence Seaway
3800 cm 3800 m 3800 km

width of a pin
1 mm   1 m   1 km

distance between Los Angeles and New York
4500 cm 4500 m 4500 km

length of a playground
50 cm   50 m   50 km

height of a professional basketball player
2 cm   2 m   2 km

greatest depth of the Pacific Ocean
11 cm   11 m   11 km

length of a large whale
20 cm   20 m   20 km

distance between two subway stops
600 cm   600 m   600 km

Changing Between Units of Length

The prefix *deci* is from *decem*, the Latin word for ten.

So __10__ decimeters = 1 meter. The symbol for decimeter is **dm.**

Complete each equation.

_____ cm = 1 m   _____ mm = 1 cm   _____ dm = 1 m

_____ mm = 1 m   _____ cm = 1 dm

To change from centimeters to millimeters, multiply by _____.

2 cm = _____ mm
5 cm = _____ mm

To change from millimeters to centimeters, divide by _____.

60 mm = _____ cm
40 mm = _____ cm

To change from decimeters to centimeters, multiply by _____.

3 dm = _____ cm
5 dm = _____ cm

To change from centimeters to decimeters, _____ by _____.

320 cm = _____ dm
670 cm = _____ dm

To change from meters to decimeters, _____ by _____.

5 m = _____ dm
12 m = _____ dm

To change from decimeters to meters, _____ by _____.

480 dm = _____ m
230 dm = _____ m

To change from meters to centimeters, multiply by _____.

7 m = _____ cm
25 m = _____ cm

To change from centimeters to meters, _____ by _____.

300 cm = _____ m
1300 cm = _____ m

To change from meters to millimeters, _____ by _____.

2 m = _____ mm

To change from millimeters to meters, _____ by _____.

7000 mm = _____ m

My millimeters are too small to distinguish!

Comparing Measurements

Change the units of one of the measurements so that both measurements are expressed in the same units. The first one is an example.

Janet's pencil is 11 centimeters long. Jose's pencil is 1 decimeter long. Who has the longer pencil? *Janet's pencil is 11 cm long.* *Jose's pencil is 1 dm, which is 10 cm.* <u>*Janet's pencil is longer.*</u>	Leonard and Darla play basketball. Leonard is 2 meters tall. Darla is 180 centimeters tall. Who is taller? _____
Chayo's work bench is 3 meters long. Ned's work bench is 370 centimeters long. Which bench is longer? _____	The Red Maple tree grows to an average height of 30.5 meters. The Silver Maple grows to an average of about 3800 centimeters. Which tree is taller, on average? _____
The length of the Nile River is about 6693 kilometers. The length of the Amazon River is about 6 436 000 meters. Which river is longer? _____	The average depth of the Pacific Ocean is 41 880 decimeters. The average depth of the Atlantic Ocean is 3735 meters. Which ocean is deeper, on average? _____
Tina's thumbnail is 1.2 centimeters long. Anita's is 16 millimeters long. Whose thumbnail is longer? _____	The deepest part of the Pacific Ocean is 11 033 meters below sea level. The deepest part of the Atlantic Ocean is 864 800 centimeters below sea level. Which ocean is deeper? _____

Decimal Parts of a Meter

Use the drawing of a meterstick below to help you fill in the blank.

[meterstick marked 1 meter]

There are _____ centimeters in a meter.

Match each centimeter measurement with its fractional measurement in meters. Then match each fractional measurement with its decimal measurement. One example is given.

1 cm	$\frac{9}{100}$ of a meter	0.09 m
9 cm	$\frac{5}{100}$ of a meter	0.12 m
12 cm	$\frac{52}{100}$ of a meter	0.99 m
5 cm	$\frac{99}{100}$ of a meter	0.01 m
52 cm	$\frac{1}{100}$ of a meter	0.05 m
99 cm	$\frac{12}{100}$ of a meter	0.52 m

Now you will work with lengths that are longer than 1 meter. Complete the table below. The first row is an example.

m plus cm	m	cm
1 m plus 16 cm	1.16 m	116 cm
		224 cm
		609 cm
	8.49 m	
5 m plus 26 cm		

Measure each length, then record each length in three different ways. Use objects in your classroom.

Object	m plus cm	m	cm
length of a table			
width of a window			
length of the chalkboard			

Team Project #5: Greater Than 1 Meter

With a partner, help each other record these measurements, in three different ways.

I can reach as high as _____ m plus _____ cm.
_____ m
_____ cm

My partner can reach as high as _____ m plus _____ cm.
_____ m
_____ cm

My height is _____ m plus _____ cm.
_____ m
_____ cm

My partner's height is _____ m plus _____ cm.
_____ m
_____ cm

My arm span is _____ m plus _____ cm.
_____ m
_____ cm

My partner's arm span is _____ m plus _____ cm.
_____ m
_____ cm

My longest step is _____ m plus _____ cm.
_____ m
_____ cm

My partner's longest step is _____ m plus _____ cm.
_____ m
_____ cm

Practice Test

Write the unit of measurement that makes the most sense (**millimeter, centimeter, meter,** or **kilometer**) for measuring each length.

length of a soccer field _____

length of the Red Sea _____

length of a pen _____

Measure the length of each object pictured to the nearest centimeter.

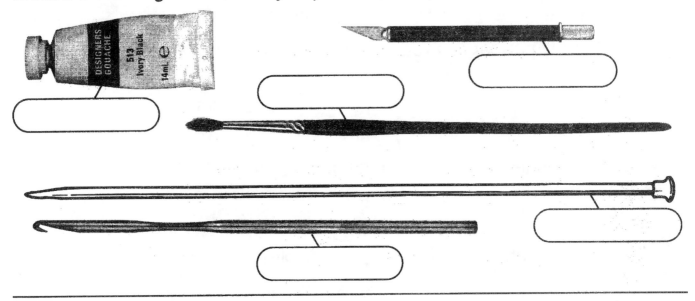

Measure the length of each line segment to the nearest millimeter.

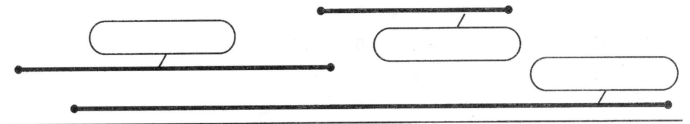

Measure the length of each line segment or the inside diameter of each circle. Record your answer to the nearest tenth of a centimeter.

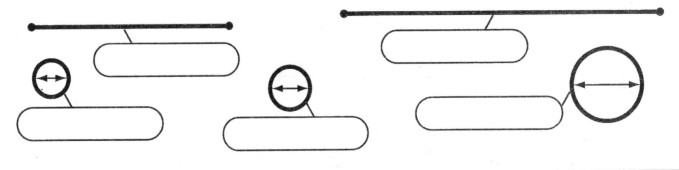

Draw a line segment for each: C ————————————— D

2 times the length of $\overline{CD}$. . . . .

1.8 cm longer than $\overline{CD}$ . . . .

First estimate and then measure
each curve to the nearest centimeter.

estimate _____   estimate _____

measurement _____   measurement _____

Solve the following problems.

Uno's longest hair is 43 millimeters long. Joe's longest hair is 4 centimeters long. Whose longest hair is longer? _____

Kyle has a pet snake that is 120 millimeters long. How many centimeters is this? _____

Ted is 1.5 meters tall. Rosie is 160 centimeters tall. Who is taller? _____

The length of the continent of Africa from north to south is 8045 kilometers. How many meters is this? _____

Fill in the blanks.

37 cm = _____ mm   7 m = _____ cm   40 mm = _____ cm

600 cm = _____ m   80 cm = _____ dm   2 mm = _____ cm

For each pair, circle the longer measurement.

6 mm or 0.5 cm   180 cm or 2 m   1 km or 100 m

Circle the best estimate of each length.

length of a van

4 cm   4 m   4 km

thickness of a piece of cloth

2 mm   2 cm   2 m

length of a pencil

15 mm   15 cm   15 m

distance from Toronto to Dallas

1800 cm   1800 m   1800 km

Projects

1. Measure and record the length of your broad jump and your running broad jump. Try to beat your own record in each. Record your results in two different ways using metric units. For example, record a jump of 1 m plus 50 cm as 1.5 m and 150 cm.

2. Figure out how many times you would have to walk the length of your block or the school yard in order to walk a kilometer. Explain how you figured this out and draw a diagram to help support your conclusion.

3. Perform the following science experiment:
   • Plant six seeds, each in a different cup.
   • Choose two different locations and place half the cups in each location.
   • Measure and record the growth of the seeds every week.
   Write about your conclusions.

4. Measure the length of your shadow every hour on a sunny day. Use your measurements to make a table to help you tell time using your shadow. Explain why this table might not work at other times of the year.

5. Measure the length, width, and height of your room. Then find the length and height of at least four different dinosaurs, and decide if each dinosaur would fit in your room. Justify your answers.

6. Ask five adults about their jobs. Find out if they ever use metric measurements. Report on your findings.

7. Interview at least two adults who have lived in other countries besides your own. Ask them about when and where they measured distances in that country and what units they used. Report on the situations they describe.

Measuring Tools

Use the rulers and the tape measures on this page to do the problems in this book. After each use, tuck them safely in your book in case you need to use them again.

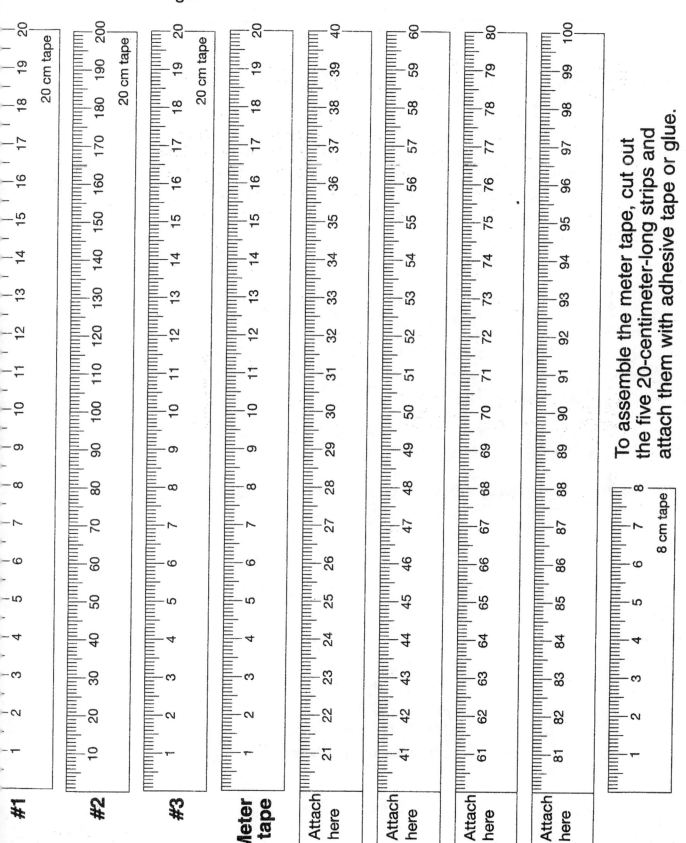

45

Key to Metric Measurement®

Book 1: *Metric Units of Length*
Book 2: *Measuring Length and Perimeter Using Metric Units*
Book 3: *Finding Area and Volume Using Metric Units*
Book 4: *Metric Units for Mass, Capacity, Temperature, and Time*
Answers and Notes for Books 1-4

Also Available

Key to Fractions®
Key to Decimals®
Key to Percents®
Key to Algebra®
Key to Geometry®
Key to Measurement®

ISBN 1-55953-325-0